See the Wi

by Linda B. Ross

Look at what the wind can do.
The wind can make kites fly.
Look at the kites go!

The wind can blow
the leaves around.
It can break an umbrella, too.

We can't see the wind.
But we see what the wind
can do.

The wind makes the boat go fast. Look how fast it goes!

The wind blows the clothes on the line.
It helps the clothes to dry.

The wind can blow
the clouds away, too.
We see the sun come out!

Then the wind stops.

The leaves stop blowing around.

The boat stops, too.

Will the wind come back?
Yes, it will!